•••

This is a very simple book.
However, that's the whole point!
It's going to show you, in SUPER
SIMPLE words, what this guy:

 Sir Isaac
Newton...

... Taught the world.
Take in mind that before Newton, people didn't
know any of this!
Yet once we explain it to you you'll be like "really?
But it's so simple!" And THAT is what makes
Newton a genius, my friend. Enjoy!

First Law

In Isaac Newton's Words:

" Every body persists in its state of being at rest or of moving uniformly straight forward, except insofar as it is compelled to change its state by force impressed."

In Simple English:

"Things don't move unless moved. Things don't stop moving unless stopped"

Explained Further. The FIRST thing to know is:

If you have an object, say a car, and it's moving at 100 mi/hr going North, the only way it will *ever* stop moving, or *even slow down* a little, or ever go anywhere but North, is if a force *acts to stop* it or change it's course.

So Why Don't I Notice This Every Day?

Normally, we don't notice this very much here on Earth because the air's friction, the friction of the road on the wheels, and other forces, even gravity, will always be pulling on the car to stop it. The car will *always* continue traveling, and *in the same direction* unless any of these forces, or any other force, changes that.

BUT, it's important to know that even here on Earth, the car will ONLY slow down TO THE DEGREE that friction, gravity etc. do act on it.

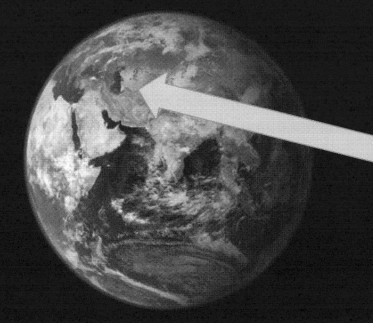

Image: NASA/GSFC

You are
here

Meaning,
here!

Frames of Reference

To understand frames of reference better, imagine a girl sitting in a moving train.

Although she is moving with the train, she can have a cup of orange juice in her hand and it won't spill at all. How is that?

Because the orange juice in the cup is moving exactly like she is, within the train, so it never goes faster or slower than anything in the train. That is why if we forget about the train, and just look at her and the cup of juice, we will observe them as standing still, and not moving at all. Both are true, she's moving in the world with the train and she is standing still within the train.

The train is moving but so is the girl and the cup of orange juice. None is moving faster or slower or in a different direction.

Seen From Outside The Train

Nothing is moving.

Seen From Inside The Train

For Example...

An aerodynamic car (one that was made to avoid as much friction from the air as possible) will, under the same conditions, stop slower than another car that is not aerodynamic (supposing nobody steps on the breaks of either car!). That's because they both are being stopped by the air, but one car is being stopped less than the other!

Car Movement

Air

Car Movement

Air (most air
Smoothly goes up)

The Second Thing To Know Is:

If an object wasn't moving, it would just never move unless a force moves it.

Although all objects are moving in the universe (for example, a person standing anywhere on Earth is moving because the Earth is moving), we can notice this law if we only look at a frame of reference (so, don't look at the Earth with the person in it, just look at the person from inside the room where she stands).

What this means, in very simple English, is that if you DON'T take into account that the Earth is moving, if you look at yourself just taking into account what you see around you (your room, the grass or wherever you are right now) and DON'T MOVE and NOTHING MOVES YOU, you will stay "not moving" (relative to your room) forever... until you decide to move.

CAN YOU BELIEVE IT?

That's it! You already undertand the first Law of Newton! CONGRATULATIONS!

IT WAS REALLY THAT EASY!

Now continue believing in yourself and don't let anything ever discourage you from knowing that you can learn physics. It's NOT hard, so don't let anyone scare you!

Second Law

In Isaac Newton's Words:

"*The change of momentum of a body is proportional to the impulse impressed on the body, and happens along the straight line on which that impulse is impressed.*"

OR:

"In an inertial frame of reference, the vector <u>sum</u> of the <u>forces</u> **F** on an object is equal to the <u>mass</u> m of that object multiplied by the <u>acceleration</u> **a** of the object: $\mathbf{F} = m\mathbf{a}$". (NOTE: It is assumed here that the mass m is constant)

In Simple English

"You know how much it hurts to be hit by a baseball? It depends on two things: 1 - how much the baseball weighs (OK, I said "weighs", because here on Earth things "weigh"; I really meant its mass though!) 2- how quickly that baseball is moving"

From this we also notice that the rate of change of momentum of anything is directly proportional to the force applied. In other words, if you hit a ball *hard*, it will go *fast*. If you hit it *softly* it will go *slowly*.

The same idea applies to the ball hitting you. If it hits you with a lot of force (and we just talked about how to calculate before) it will hurt more than with little force. You might or not move when it hits you, but the only way you can not move is if you "stand your ground", meaning if you use your own force to stop yourself from moving once it hits you, in which case all the force with which it hit you will be used to make it hurt your arm or wherever it hit you even more!

Explained Further

A baseball that weighs 2 lb going at 2 yards every second will hit you harder than the same baseball going at 1 yard per second, because it's going slower. Also, it will hit you harder than a baseball that weighs just 1 lb even if it, too, is going at 2 yards every second because it weighs more.

If you want to know the exact force with which it will hit you, just do the following:

Multiply the mass of the baseball (2lb) times the acceleration (just call it speed for now) (2 yards every second). That is equal to the force (in this case 4)

In the Lab

That's all the theory you need to know on this law!

However, if you want to solve some physics problems, it would be good to use the same units scientists worldwide use (meters/second) and use kg for mass. Use "Newtons" for the unit in which your force, which is the answer to multiplying the speed times the mass, will be in.

Acceleration Vs. Just Speed

Speed is how fast an object is moving.

Sometimes the baseball will be accelerating, meaning it's not traveling just at 2 m/second but actually it's traveling faster and faster every second (for example, if it's falling down). In that case, instead of using a constant speed, you use an acceleration in your formula.

Calculating that is very easy, but to keep things simple, just follow this link (or type it in your browser for paperbacks) if you want to get into that now:

https://www.wikihow.com/Calculate-Acceleration

Congratulations!

That's it! You already undertand the first Law of Newton AND the Second Law of Newton! CONGRATULATIONS!

IT WAS REALLY THAT EASY!

Now continue believing in yourself and don't let anything ever discourage you from knowing that you can learn physics. It's NOT hard, so don't let anyone scare you!

Third Law

In Newton's Words:

"To every action there is always opposed an equal reaction: or the mutual actions of two bodies upon each other are always equal, and directed to contrary parts"

In Simple English:
"Basketballs bounce because when they hit the floor, they are hit back with the same amount of force that they hit the floor. That's also why if you hit the floor it hurts YOU and it doesn't only 'hurt' the floor!"

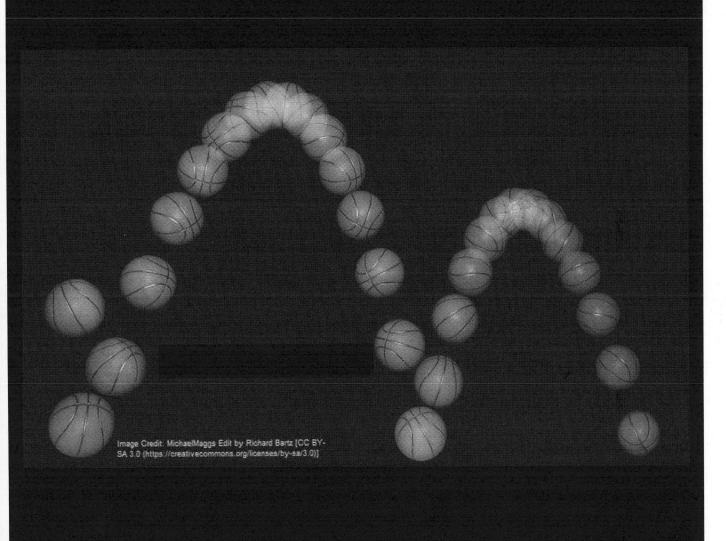

"For Every Action There's An Equal But Opposite Reaction"

Think about it this way:

A basketball hits the floor. All the force with which the basketball hit the floor (and you can calculate that force with the Second Law), has to go SOMEWHERE, right? It can't just disappear! So, where does that force go? Right back where it came from into the ball! That's why the ball bounces!

Examples:

1) A person swims. She is pushing the water backwards. Therefore, the water is pushing her forwards.
2) A basketball bounces. It is pushing the force downwards, so the floor pushes it upwards.
3) A person walks. She is stepping on the floor and pushing it down and backwards, to be pushed by the floor forward and slightly upwards on every step.

Congratulations!

You understand all 3 Laws of Newton!

Wasn't that incredibly easy?

Yes, you can go deeper and deeper, but you already REALLY have A LOT, and it was SO EASY! CONTINUE ENJOYING PHYSICS AND DON'T EVER BELIEVE IT'S TOO HARD FOR YOU!

Can you do me a Favor?

If you liked this book, could you please give it a good review on Amazon? THANKS!

Made in the USA
Las Vegas, NV
24 October 2024